1 MONTH OF
FREE
READING

at

www.ForgottenBooks.com

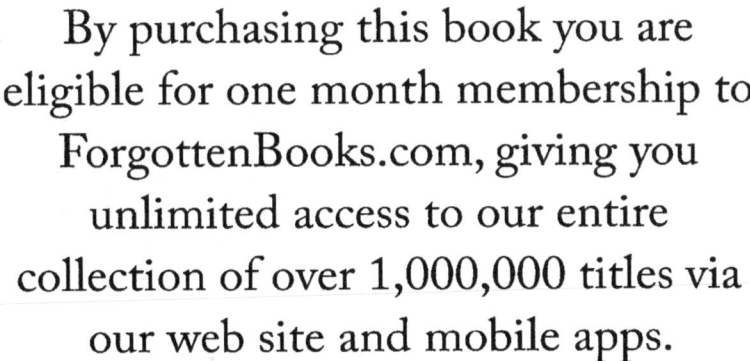

By purchasing this book you are eligible for one month membership to ForgottenBooks.com, giving you unlimited access to our entire collection of over 1,000,000 titles via our web site and mobile apps.

To claim your free month visit:

www.forgottenbooks.com/free59951

ISBN 978-0-266-16144-8
PIBN 10059951

This book is a reproduction of an important historical work. Forgotten Books uses
state-of-the-art technology to digitally reconstruct the work, preserving the original format
whilst repairing imperfections present in the aged copy. In rare cases, an imperfection in
the original, such as a blemish or missing page, may be replicated in our edition. We do,
however, repair the vast majority of imperfections successfully; any imperfections that
remain are intentionally left to preserve the state of such historical works.

Bulletin No. 254

Series { A, Economic Geology, 49
{ B, Descriptive Geology, 61

DEPARTMENT OF THE INTERIOR

UNITED STATES GEOLOGICAL SURVEY

CHARLES D. WALCOTT, DIRECTOR

REPORT OF PROGRESS

IN THE

GEOLOGICAL RESURVEY OF THE CRIPPLE CREEK DISTRICT, COLORADO

Walter W. Bradley.

BY

WALDEMAR LINDGREN AND FREDERICK LESLIE RANSOME

WASHINGTON
GOVERNMENT PRINTING OFFICE
1904

CONTENTS.

3

LETTER OF TRANSMITTAL.

DEPARTMENT OF THE INTERIOR,
UNITED STATES GEOLOGICAL SURVEY,
Washington, D. C., November 3, 1904.

SIR: I have the honor to transmit herewith the manuscript of a preliminary report on a "Resurvey of the Cripple Creek district of Colorado," by Messrs. Waldemar Lindgren and F. L. Ransome, under my general supervision.

Although only ten years had elapsed since a careful study of the geology of this district was made by Messrs. Cross and Penrose, the people of Colorado had been so strongly impressed with the economic importance of a scientific examination of the ground opened by mining operations during that period, that they urgently requested this resurvey and have materially assisted in its execution.

The present report, made in advance of the final laboratory examinations, is a summary of those facts that bear upon the economic development of the region and are of immediate importance to the miners. I therefore request that it be published with the least possible delay, that its results may be promptly available to all interested in the region.

Very respectfully,

S. F. EMMONS,
Geologist in Charge Section of Metalliferous Deposits.

Hon. CHARLES D. WALCOTT,
Director United States Geological Survey.

REPORT OF PROGRESS IN THE GEOLOGICAL RESURVEY OF THE CRIPPLE CREEK DISTRICT, COLORADO.

By WALDEMAR LINDGREN and FREDERICK LESLIE RANSOME.

INTRODUCTION.

The Cripple Creek gold deposits were discovered in 1891. Shortly afterwards, in 1894, an examination of the new district was undertaken by the United States Geological Survey, Mr. Whitman Cross having charge of geology and petrography and Mr. R. A. F. Penrose, jr., undertaking the examination of the mines. Their report, accompanied by a geological map, was published in the Sixteenth Annual Report of .the Geological Survey, Part II, pages 1–207, and has for the last ten years served as a useful and accurate geological guide to mining operations.

With the astonishingly rapid development of the Cripple Creek mines the opportunities for geological study multiplied, revealing great complexity of vein phenomena and stimulating a desire for further investigation, particularly with a view of obtaining some evidence as to the persistence of the veins in depth. This desire found expression in a request by citizens of Colorado for a reexamination of the district by the United States Geological Survey and in an offer of cooperation, whereby the cost would be equally divided between the State of Colorado and the National Survey. The necessary amount having been subscribed in Cripple Creek, Colorado Springs, and Denver, the State contribution was put in the hands of Mr. John Wellington Finch, State geologist of Colorado, and by him disbursed. The cordial thanks of the geologists in charge of the work are due to Mr. Finch for his hearty and efficient cooperation.

The reexamination began with a thorough revision of the topographic map of Cripple Creek by Mr. R. T. Evans, Mr. E. M. Douglas in charge. This involved a great deal of painstaking work, nearly every prospect hole being located, as well as all shafts and tunnels. The new map is on the scale of 1:19,495, or about 2⅓ inches to the mile, and includes practically the same area as the old map, a small strip only being added on the western side, so that the total area mapped is about 40 square miles. The small developments outside of this area did not seem to justify further extension of the boundaries. Contours are 50 feet apart, and a numbered list of 324 mines is given on the margin of the sheet. A line of accurate levels was run to Cripple Creek from Colorado Springs, thus settling the conflicting data of the different railroads.

The geological and mining work was undertaken jointly by the authors of this preliminary report. The examination began in June, 1903, and, with some interruptions, due to various causes, the field work was concluded in April, 1904. Practically every accessible mine in the district was examined in greater or less detail. Mr. L. C. Graton served as assistant throughout this time, actively participating in all branches of the work. Messrs. A. M. Rock and J. Bruce also rendered very efficient aid as draftsmen.

PRODUCTION.

Though situated close to the centers of population in Colorado and in an easily accessible region, the gold deposits of Cripple Creek were not discovered until 1891. To a great extent the lateness of the discovery was due to the extremely inconspicuous character of the vein croppings and to the equally inconspicuous appearance of the dark-brown, powdery gold set free by the oxidation of tellurides. As soon as the true character of the veins was ascertained the development of the district proceeded rapidly. In 1894, when the first survey was made, the production was a little less than $3,000,000, but the next year this amount was more than doubled, and in 1900 the maximum production of a little over $18,000,000 was attained. In 1901 and 1902 the production declined slowly, and dropped the next year to $13,000,000. The sudden decrease in 1903 was to some extent brought about by the impoverishment of several mines, but the labor troubles of that year had also much to do with it. From August, 1903, to the summer of 1904 many mines experienced more or less difficulty from this cause. It is probable, however, that the output for the current year will show a considerable increase over that of 1903.

Production of the Cripple Creek district according to the reports of the Director of the Mint.[a]

Year.	Gold.	Silver.
		Fine ounces.
1891	$449	
1892	583,010	
1893	2,010,367	5,019
1894	2,908,702	25,900
1895	6,879,137	70,448
1896	7,512,911	60,864
1897	10,139,709	57,297
1898	13,507,244	68,195
1899	· 15,658,254	82,520
1900	18,073,539	80,166
1901	17,261,579	90,884
1902	16,912,783	62,690
1903	12,967,338	42,210
Total	124,415,022	646,193

The total dividends can not be ascertained on account of the many individuals and small companies operating in the district. The dividends of the larger companies, it is stated, amounted to $32,752,000 to the end of 1903. In that year it is reported that $1,716,000 was paid by fourteen mines, Portland, Strong, and Stratton's Independence leading, with $360,000, $300,000, and $250,000, respectively.

MINING AND METALLURGY.

At the present time there are about 300 mines in the district, though many of these are idle and others are consolidated into larger properties. The number of shafts 1,000 feet or more in depth is about 22; of these the Lillie is the deepest, having attained 1,500 feet. Comparatively few of the mines are pumping, the district being drained by tunnels, as will be described in a following paragraph. The motive power for hoisting is ordinarily steam, electric power being used only in smaller shafts. The mining methods employed present few features of particular interest. The width stoped ranges from 3 to 50 feet or more. The stopes are sometimes filled, but are often left open after the broken rock between levels has been drawn off. Operations are facilitated by the great hardness of the rock, stopes 200 feet or more in height sometimes standing for years. To an annual production of $18,000,000 (1900) corresponded a maximum output of about 600,000 tons. The ores are not adapted to concentrating by ordinary means.

---

[a] The figures given in the Mint reports are considerably lower than those usually quoted in mining journals and popular descriptions of the district.

Hand sorting and washing, the latter in order to separate the fines, are the methods employed. More care is now taken than formerly, but at many places there is still room for improvement.

The district contains many large mines operated by strong companies, but the system of leasing to individuals and small companies, with royalty charges of 15 to 25 per cent of the net output, remains a conspicuous feature. At the present time the mines of the northwestern and northeastern parts of the district are only slightly productive. Among the formerly highly productive mines in this section may be mentioned the Victor, Isabella, Wild Horse, Damon, Logan, Orpha May, Pharmacist, C. O. D., Gold King, and Anchoria-Leland. The southwest quarter of the district contains the active and important El Paso, Mary McKinney, and Elkton mines, but the maximum output during the last few years has come from the southeastern portion, within which are the Portland, Gold Coin, Ajax, Strong, Golden Cycle, Vindicator, and Last Dollar mines. The Portland mine has for some time been the heaviest producer in tonnage as well as in value. Its output is from 60,000 to 90,000 tons per annum. In 1903 this mine produced $2,600,000.

The metallurgical history of the camp is interesting. Beginning with local amalgamating mills, the practice soon changed to local chlorination and cyanide plants. It was soon found, however, that better situations would be found in the valleys, and at present a very large part of the tonnage is shipped to chlorination and cyanide works at Colorado City and Florence. About a sixth of the output, comprising the higher grade ores, is sent to smelting works at Denver or Pueblo. A still smaller amount is treated in a local chlorination plant near Victor. While in former years the practice leaned more toward the cyanide process, the bulk of the ore is now treated in large chlorination mills with automatic roasters and revolving barrels. It has always been found necessary to roast all except the completely oxidized ores. At the present time there are in the district two small mills in which the cyanide process is used without previous roasting, and which are thus enabled to work ores containing as little as $5 a ton in gold. The extent of these low-grade ores is not yet fully demonstrated. Regarding the value of the ore, see page 21.

Treatment charges at the mills fluctuate somewhat, but usually range from $7 to $13 a ton, according to the tenor of the ore; in the early part of 1904 the prices were reduced, it is reported, and ranged from $5.50 to $10 per ton. Recently they have again been increased. On the whole the mining and milling expenses are very high at Cripple Creek compared with those of other localities—western Australia, for example, where similar ores and conditions prevail. Few of the large mines seem to have reduced total expenses below $16 per ton.

TOPOGRAPHY.

The mines are situated in a group of bare, rounded hills forming part of the high plateau extending southwestward from Pikes Peak, and are only about 10 miles distant from that prominent landmark. The elevations range from 9,000 to nearly 11,000 feet above sea, the highest point in the district being Trachyte Mountain (10,863 feet). Bull Hill and Bull Cliff are but slightly lower. The drainage is chiefly southward toward the Arkansas River. The district contains two important towns. Cripple Creek is situated on the northwest side of the producing area, while Victor, 3 miles distant, lies on the southwest edge of the same area. Two railroads connect the district with Colorado Springs, the Colorado Midland circling around the north side of Pikes Peak, while the Short Line descends to the valley along the picturesque eastern slope of the same mountain. The Florence and Cripple Creek Railroad runs southward to Florence in the Arkansas Valley. An excellent system of electric-car lines connects the towns with all the important mines.

GENERAL GEOLOGY.

FIRST GEOLOGICAL SURVEY OF THE DISTRICT.

When Mr. Whitman Cross made his careful study of the geology of the Cripple Creek district, ten years ago, mining had barely begun and the various hills were not, as now, perforated by deep underground workings. That his work has in general stood the test of subsequent underground exploration and continues to be highly regarded in the district is convincing proof of its high quality. Later workers, however they may modify or amplify his results, should acknowledge their debt to the pioneer who first deciphered the history of this volcanic district. The account of general geology, as given by Cross, may be very briefly summarized as follows:

The Cripple Creek hills lie near the eastern border of an elevated and much dissected plateau which slopes gently westward for 40 miles, from the southern end of the Colorado Range, dominated by Pikes Peak, to the relatively low hills connecting the Mosquito and Sangre de Cristo ranges. The prevailing rocks of this plateau are granites, gneisses, and schists. The granites inclose masses of Algonkian quartzite and are therefore post-Archean. But they are older than the only Cambrian sediments known in Colorado, and on the Cripple Creek map have been indicated as Algonkian. During Tertiary time volcanic eruptions broke through these ancient rocks at several points and piled tuffs, breccias, and lavas upon the uneven surface of the plateau. The eruptive rocks of the Cripple Creek district are the products of one of the smaller isolated volcanic centers of this period, a center characterized by the eruption of phonolite, which does not occur elsewhere in this general region.

The most voluminous products of the Cripple Creek volcano now preserved are tuffs and breccias. They occupy a rudely elliptical area in the center of the district, about 5 miles long in a northwest-southeast direction and about 3 miles wide. According to Cross these breccias and tuffs rest in part upon an earlier flow of andesite, but mainly upon an unevenly eroded surface of the granites and schists, although along the southwest edge of the area the contact was found to be so steep as "to support the idea that the central vent or vents of the volcano were adjacent to this line." The breccia is much indurated and altered, but was thought by Cross to consist mainly of andesitic fragments, although it was recognized that fragments of phonolite are locally abundant. The most characteristic massive rock of the Cripple Creek volcano is phonolite, which was erupted at several periods and more abundantly than any other type. It occurs as dikes and masses, not only in the breccia but in the surrounding granitic rocks.

The general succession of igneous rocks, according to Cross, is as follows: The earliest rocks were andesites containing some orthoclase. Then came a series of allied phonolitic rocks, rich in alkalies and moderately rich in silica, together with some andesites. Among them are trachytic phonolite, nepheline-syenite, syenite-porphyry, phonolite, mica-andesite, and pyroxene-andesite. Phonolite was erupted at several periods. The nepheline-syenite he considered as probably younger than the trachytic phonolite. At the close were intruded a small number of narrow dikes of basic rocks, the so-called basalts, which contrast very markedly with the phonolite.

MODIFICATIONS OF EARLIER RESULTS.

In the course of the present investigation the geology of the district has been entirely remapped upon the carefully revised topographic base. The granites, gneisses, and schists have been differentiated and outlined in greater detail than was practicable in the earlier investigation.. The oldest rocks in the district are muscovite- and fibrolite-schists. These are closely associated with the fine-grained granitic gneisses such as underlie most of the town of Cripple Creek. This gneiss, in the earlier report, was mapped partly as schist and partly as granite. Both gneiss and schist are cut by a reddish granite which occupies a considerable area extending from Anaconda westward beyond the limits of the area studied. This granite is well exposed along Cripple Creek in the vicinity of Mound.

A second type of granite distinguished and mapped is the coarsely porphyritic rock referred to by Cross as the Pikes Peak type of granite. This rock occupies over half of the district and is the prevailing type along the northern, eastern, and southern borders of the area. It is well exposed on Squaw Mountain and is the granite of the El Paso, Elkton, Ajax, Portland, Independence, and Gold Coin mines. Of these two kinds of granite the Cripple Creek variety is probably the younger. The Cripple Creek granite, the gneiss, and the schist together form a wedge-shaped area projecting into the Pikes Peak granite from the west. The center of volcanic disturbance practically coincides with the point of this wedge.

The present investigation indicates some necessary modifications of the earlier report in the way of stronger emphasis on the intimate genetic relationship of the rocks. The "phonolite," "nepheline-syenite," "trachytic phonolite," "syenite-porphyry," and "andesites" of Cross are all very closely related and have been found to be in most cases connected by intermediate types. They are clearly all slightly divergent eruptive facies of one general magma characterized chemically by containing from 9 to 15 per cent of potash and soda, the soda being always somewhat higher than the potash, particularly when the comparison is made by molecular ratios. None of the massive rocks can properly be called andesite, and although it can not be affirmed that andesitic fragments are entirely absent from the usually much altered volcanic breccia, the term "andesitic breccia" does not seem applicable to this formation as a whole. It would be more accurate to describe it as a phonolitic breccia, although in places it consists chiefly of particles of the older rocks through which the Tertiary eruptives broke.

None of the massive rocks erupted from the Cripple Creek volcanic center and now preserved in the district show any evidence of having been surface flows. They are for the most part intrusive porphyries, ranging in texture, however, from the granular so-called nepheline-syenite near the town of Independence to the nearly aphanitic phonolite of the smaller dikes and sheets. Most of them will come under the designations phonolite, trachytic phonolite, trachydolerite, and alkali-syenite. The extensive underground workings show that the "nepheline-syenite" does not cut the "trachytic phonolite," but that the two rocks represent textural and, to some extent, mineralogical facies of the same mass, while the trachytic phonolite in turn may pass into phonolite. The trachytic phonolite is in some instances cut by dikes of phonolite, showing, as Cross has already pointed out, that the phonolitic intrusions were not all synchronous.

While it is undoubtedly true that much of the breccia in the northeastern part of the volcanic area rests upon a very uneven surface of granite, gneiss, and schist, the results of field work during the last season, favored by deep workings not in existence when the district was originally surveyed, have emphasized the fact that the breccia lying southwest of a general northwest-southeast line drawn through

Big Bull Mountain and Gold Hill occupies a chasm of profound depth in the fundamental rocks of the region. From the Conundrum mine on the western slope of Gold Hill to Stratton's Independence mine on the south slope of Battle Mountain the contact plunges steeply down, with dips ranging in general from 70° to vertical. In some instances the granite walls of this chasm actually overhang the breccia. It is probable that this entire southwest contact represents a part of the wall of the great pit formed by the volcanic explosions that produced the breccia. It is further probable that an arm or branch of this volcanic abyss, now filled with breccia and intrusive rocks, extends northeastward past Bull Cliff and the town of Altman.

ECONOMIC GEOLOGY.

EARLIER WORK.

To the excellent work of Mr. R. A. F. Penrose, jr., apply statements similar to those made in the discussion of the purely geological branch of the subject. Few shafts had then attained a depth of 400 feet, and most of the exposures were masked by surface oxidation. It would be surprising, in view of the facilities created by the later development of hundreds of mines, if a subsequent investigation should not bring out some slight modifications of earlier results.

EXTENT OF PRODUCTIVE TERRITORY.

There is nothing in the history of the district since 1894 warranting any extension of the bounds of the productive territory as then known. Now, as then, a circle of 3 miles radius described from the summit of Gold Hill would include all deposits of known or prospective value, while the really important mines would be embraced by a circle of about half that radius, with its center near the summit of Raven Hill. That scattered deposits of greater or less value may be found in outlying portions of the district is by no means improbable. But the close dependence of the typical Cripple Creek ores upon the main volcanic center, and the consequent remarkable compactness of the gold-bearing area, are features highly characteristic of the district and are likely always to remain so.

The greater part of the ore has undoubtedly come from the central area of breccia, particularly from that part of this area in which the breccia extends to great depth. Very productive ore bodies have been found, however, in the granitic rocks, usually within 1,000 feet of the steep contact limiting the breccia on the southwest. The important Beacon Hill mines, with ore bodies nearly three-quarters of a mile from this contact, are exceptional, and are probably genetically connected with the intrusive mass of phonolite forming the core of the hill.

UNDERGROUND DEVELOPMENT.

At the time of the earlier survey the deepest shafts, those of the Moose, Pharmacist, and Anna Lee mines, were down only about 400 feet, while few of the other mines were over 200 feet in depth. Many subsequently prominent mines were then mere prospects or had not been located.

The deepest shaft at present is the Lillie, which is over 1,500 feet deep, although the Stratton's Independence shaft, 1,400 feet deep, has the lowest sump in the district. The American Eagle shaft is nearly as deep as the Lillie, while there are about twenty other shafts over 1,000 feet in depth, and at least 100 shafts deeper than the deepest workings existing in 1894. As regards absolute elevations, the Gold Hill shafts are scarcely down to a level of 9,000 feet above sea; the Elkton, El Paso, and Lillie shafts descend to 8,750 feet; Stratton's Independence reaches the lowest level at 8,450 feet; while the Gold Coin shaft, at 8,550 feet, is of interest from the fact that the deepest ore shoot in the district is now being stoped from its twelfth level.

The amount of drifting and crosscutting accomplished since the earlier survey is more than commensurate with the increased number and depth of the shafts, and the district is further intersected in various directions and at different levels by two long tunnels run for drainage purposes and by a dozen or more extensive adits, many of which have their portals in the granitic rocks and extend well into the central part of the breccia area.

BRIEF REVIEW OF THE MINES.

The productive district, as stated above, is practically covered by the area of a circle 3½ miles in diameter. The center of this circle would be located halfway between Raven Hill and Bull Hill, and the towns of Cripple Creek, Victor, and Cameron would be situated on its periphery. A very few mines—notably the Galena and the Fluorine—and many prospects lie outside of this area.

The culminating points of the district are found in a ridge of high and bare hills that extends in a northwest–southeast direction and divides the waters flowing into Cripple Creek and Wilson Creek on the southwest from those joining Spring Creek and Grassy Creek on the north. From northwest to southeast the following hills mark this divide: Mineral Hill, Carbonate Hill, and Tenderfoot Hill, north or northeast of Cripple Creek; Globe Hill, Ironclad Hill, and Bull Hill, the latter being near the center of the district and equidistant from Cripple Creek and Victor; the ridge is continued by Bull Cliff and Big Bull Mountain, the latter, really outside of the productive area, being the highest point in this dividing range of hills. Its elevation is 10,826 feet. Three long spurs project to the southwest from the dividing range separating the deep trenches of Cripple Creek, Squaw Gulch, Arequa Gulch, and Wilson Creek; the first, called Gold Hill, rises directly east of Cripple Creek; the second is Raven Hill, being continued to the southwest by the lower spur of Guyot and Beacon hills; the third is Battle Mountain, continued by the almost equally high salient of Squaw Mountain.

The important mines are situated in this region of sharply accentuated topography. As has been several times emphasized, the volcanic area practically coincides with the hills and ridges just described and is surrounded on all sides by granitic rocks.

Globe and Ironclad hills and Gold and Raven hills consist chiefly of heavy masses of breccia, and were scenes of great activity during the early years of the district. Near Poverty Gulch, just northeast of Cripple Creek, is the Abe Lincoln, not a large mine, but still actively worked with satisfactory results. Higher up are the Gold King, with dividend records of $150,000, and the C. O. D., with a reported production of $600,000 and dividends of $150,000. Both were idle in 1904 and have attained their eighth or ninth levels.

On the summit of Globe Hill are the Stratton properties of Plymouth Rock and Globe mines, in which extensive low-grade mineralization without many sharply defined veins seems to be the rule. Adjoining is the property of the Homestake Company, including the Ironclad mine, where direct cyaniding of oxidized surface ores is now carried on in a mill erected on the property.

Gold Hill is crowned by the Anchoria-Leland mine, with a production of over $1,000,000 and dividends of $198,000. The shaft is 1,100 feet deep. The adjoining Moon-Anchor has paid dividends of $261,000, and the Half Moon (Matoa G. M. Co.) has a gross production of $650,000 to its credit, but is reported to have paid only a small amount in dividends. None of these mines is being worked at present, except on a small scale by lessees.

On the western slope is the Midget mine, actively worked at present, with a depth of 800 feet, a total production of $662,000, and dividends of $195,000. The Conundrum, in the same vicinity, is likewise worked with good results to a depth of 600 feet. The Midget, like the mines described above, follows a vein in breccia, while the Conundrum is mining on a "basalt" dike in granite, close to the contact of the breccia.

In the deep gulch between Gold Hill and Raven Hill are situated the Anaconda, Doctor-Jack Pot, and Mary McKinney mines, all working on sheeted zones forming lodes in the breccia. The Anaconda produced about $1,000,000, chiefly from upper levels, and is now idle. The Mary McKinney is one of the most successful mines worked at present in the district. Its depth is 600 feet. The Doctor-Jack Pot has $4,000,000 to its credit and likewise a handsome dividend record.

The shaft is only 700 feet deep, water having until now prohibited deeper sinking.

The breccia-granite contact is found on Guyot Hill a short distance south of the Mary McKinney. The extreme spur of Raven Hill, called Beacon Hill, is formed of an intrusion of phonolite in granite, and about this outlying volcanic center cluster a group of veins of great production and promise. On the eastern side of the hill are located the Prince Albert, Gold Dollar, and others, not active producers at present, while on the western side lie the El Paso, C. K. & N., and Old Gold mines, with their narrow but extremely rich fissure veins in granite, now actively and successfully worked.

A great number of smaller mines have been worked on veins cutting the breccia of Raven Hill. The famous Elkton mine is situated in the deep hollow between Raven Hill and Battle Mountain. It has been working on an exceptionally long vein, partly contained in breccia, partly in granite, and generally following a "basalt" dike. The production approaches $6,000,000, and the depth attained is about 900 feet, excessive water having formed a serious obstacle to deeper sinking. Dividends amount to $1,200,000. The Moose mine, situated higher up on the slope of Raven Hill, had a good ore shoot, from which $500,000 was obtained.

Continuing northwest, we soon attain the summit of Bull Hill, which affords a magnificent panorama, not only of the whole camp, but of a large part of the State of Colorado. Toward the east, and 5,000 feet lower, spread the great plains at the foot of the Rocky Mountains; westward the Sangre de Cristo, Collegiate, and Mosquito ranges—a snowy and jagged line of ramparts—define the distant horizon.

A multitude of small mines occupy the southwestern slope of Bull Hill. On the northwestern side an area of brecciated granite appears among the volcanic rocks, and in this formation is situated the Wild Horse mine. This lode, which has been worked to a depth of 1,250 feet, has produced over $1,000,000, but is now operated only by lessees. A number of smaller producers may be found on the northern slope, toward Cameron, among them the Damon, Jerry Johnson, W. P. H., and Pinnacle.

Those who have followed this description on a map will have noticed that the mines are chiefly situated on the periphery of a circular area, the central part of which, comprising the upper part of Squaw Gulch, has thus far yielded very little. Few strong veins have been met with in this part of the breccia, but, on the other hand, the developments in depth are not extensive.

On the east and southeast side of Bull Hill begins that most important belt of lodes which extends southward to Victor and includes the richest group of producers in the camp. A characteristic feature of this belt is the intrusion into the breccia of thick masses of trachytic phonolite and syenitic rocks.

With few exceptions the veins of this belt strike north-northwest. We may begin the description with the system of linked veins, 3,000 feet long, covered by the Isabella and Victor mines. The last-named mine, on the southern end of the system, is situated just below the western slope of Bull Cliff. It has been worked to a depth of over 1,000 feet, has produced about $2,200,000, and has paid dividends amounting to $1,150,000. The Isabella has attained a depth of 1,127 feet, produced $3,200,000, and paid dividends of $600,000. Both mines lost their pay shoot in depth.

The small but rich cross veins of the Empire State, Burns, Pharmacist, and Zenobia connect this vein system with that of the Stratton mines on Bull Hill. South of the Burns begins the great Vindicator vein system, traced southeasterly for a mile through the Findley, Hull City, Vindicator, Lillie, and Golden Cycle mines. The Hull City and the Lillie have each produced over $1,000,000, the Vindicator and Golden Cycle over $2,000,000 each, all with corresponding dividend records. The Lillie is deepest, having attained 1,500 feet. Next in depth is the Vindicator, 1,200 feet. All of them, except the Lillie, are still actively worked. In the whole system water has been and is still a source of trouble. The deepest mine evidently drains all the others in this vicinity.

The Stratton properties on Bull Hill, with the Logan, Orpha May, and Pikes Peak veins, on which maximum depths of 1,200 and 1,500 feet have been attained, are now worked only to a slight extent, whereas in the early days of the camp they were highly productive.

This vein system is continued southward in the Last Dollar mine, now working at a depth of 1,270 feet. The production exceeds $1,000,000. South of the Last Dollar the veins enter the Modoc ground, a mine worked for a long time and with gratifying success. The Blue Bird, an old-time producer, is situated a short distance west of the Last Dollar.

South of the Modoc is the Battle Mountain vein system, crossing from the granite into the breccia, with general northerly or north-northwesterly directions, and distinguished by heavy production and ore bodies of imposing size. None of the veins are of great length, and the whole system extends scarcely a mile along the strike of the veins. The veins can not be directly connected with others already described, though, in its general trend, the system heads toward the Dexter, Blue Bird, and Moose veins.

Beginning on the southwestern side, we first come to the Gold Coin mine, the veins of which are in granite; one of them is successfully worked at present at a depth of 1,200 feet. The total production approaches $6,000,000; the dividends paid exceed $1,000,000. North of the Gold Coin is the Ajax, working partly in veins, partly in large, irregular ore bodies in the granite. The total production is very considerable. The depth attained is 1,200 feet.

Between this and the Portland vein system, almost within the town of Victor, are the Granite, Dillon, and Dead Pine veins. They are worked at present at depths of from 800 to 1,000 feet.

The Portland vein system begins on the south at the Strong mine, now worked at a maximum depth of 900 feet, on a vein in granite that follows a "basalt" dike, which is in places accompanied by a phonolite dike. The mine is an unusually regular and profitable producer, the total dividends since 1892 amounting to $2,500,000.

The veins of Stratton's Independence run about parallel to those of the Strong, a few hundred feet eastward. They extend from the granite into the breccia, following for some distance a phonolite dike. The production of this mine amounts to over $11,000,000, with a dividend record of $4,000,000 since 1899. At present the company is leasing the various levels to tributers. From the two properties last described the vein systems continue into the Portland mine, but in the northern part of that great property are replaced by another and still richer aggregate of veins, the Captain system. The Portland is, beyond question, the most prominent mine of the Cripple Creek district. Its total production from 1894 to the end of 1903 amounted to $18,000,000, derived from 466,000 tons of ore (both in round figures), from which $4,600,000 has been paid in dividends, the remainder going to acquirement of territory, extensive milling and mining plants, and operating expenses.

Outside mining properties.—The area outside of the principal volcanic area contains very few productive properties, but it is by no means barren. A great deal of money has been spent here, usually with unsatisfactory results. Although there are many properties of merit and although much honest effort has been made in this part of the district, it has long been the favorite camping ground of concerns more or less lacking in stability.

The granite hills west and south of the city of Cripple Creek contain few prospects; phonolite dikes occur in places, but usually show little value. Along Gold Run and Arequa Gulch prospects with a little ore have been found, down to the junction with Cripple Creek, and even at isolated places below this locality. Grouse Mountain, with its phonolite cap, shows many prospects from which occasional good assays have been obtained, but neither here nor on Straub and Brind mountains has anything of permanent value been developed thus far. It is claimed that ore bodies of low grade, containing a few dollars per ton, exist.

The breccia caps of Mineral, Carbonate, and Tenderfoot hills are dotted with prospect dumps, and even shafts several hundred feet deep. Nothing of permanent value is recorded from Mineral Hill, though fairly productive placers have been worked at its southwestern base, almost in the town of Cripple Creek.

On Carbonate Hill the Elkhorn has been a small producer; on Tenderfoot Hill the Friday, Hoosier, Black Diamond, and Mollie Kathleen contribute their parts to the production. Two miles north-northwest of Cripple Creek is the Galena mine, the vein of which follows, for a part of its course, a phonolite dike in granite and has a small output to its credit. About the same distance north of the city is the small volcanic center of Copper and Rhyolite mountains. At the former the Fluorine mine has produced $160,000, and low-grade ore is now being cyanided. Prospects are found on Rhyolite Mountain, and in fact all over the flat granite country between it and Trachyte Mountain. The Lincoln mine, near Gillette, and several other prospects farther south, along a belt of phonolite dikes, have produced a little ore. It is claimed that there are low-grade veins on both sides of Bernard Creek, northwest of Gillette, in a region of granite with occasional dikes and masses of phonolite. Trachyte Mountain, southeast of Gillette, is covered by phonolite, and a little ore is occasionally found in veins at its southern foot. Some work has also been done on Cow Mountain, about 4 miles northeast of Bull Hill.

The eastern margin of the central volcanic area, east of Victor Pass and extending southward across Big Bull Mountain to Brind Mountain, has thus far failed to produce anything of importance, though well covered by prospects. A survey of these outlying parts of the district serves to emphasize strongly the remarkable concentration of deposits within the narrow limits of the central volcanic area.

CHARACTER OF THE ORES.

The characteristic feature of the Cripple Creek ores is the occurrence of the gold in combination with tellurium, chiefly as calaverite, but partly also as the more argentiferous sylvanite,[a] and probably to a minor extent as other gold, silver, and lead tellurides. The tellurides are frequently associated with auriferous and highly argentiferous tetrahedrite, with molybdenite, and occasionally with stibnite. While these minerals have not yet been closely studied, preliminary examination indicates that their contents in gold are due to an intimate mechanical mixture of tellurides. Pyrite, while widely disseminated through the country rock and of common occurrence in the fissures, is rarely sufficiently auriferous to constitute ore. Such of the pyritic ores as have been tested reveal the presence of tellurium, indicating that the ore is a mixture of pyrite and gold-silver tellurides. Galena and

---

[a] Calaverite (AuAg) Te_2: tellurium, 57.4 per cent; gold, 39.5 per cent; silver, 3.1 per cent. Sylvanite (AuAg) Te_2: tellurium, 62.1 per cent; gold, 24.5 per cent; silver, 13.4 per cent.

sphalerite occur in small quantities in many of the mines, but rarely contain enough of the precious metals to form ore. Native gold appears to be absent from the telluride ores, except as it may be set free by the oxidation of these tellurides.

The usual gangue minerals of the ores are quartz, fluorite, and dolomite. Roscoelite and rhodochrosite are also found in places. Celestite, or sulphate of strontium, while never present in large amount, frequently occurs as little acicular crystals in the quartz vugs of the lodes. Calcite occurs interstitially in much of the breccia near the ore bodies, but is rarely found in distinct crystalline form with the ore minerals. Secondary potassium feldspar is common in the ores; it is especially abundant in the ores inclosed in granite, particularly those in the Pikes Peak type. This feldspar has the composition of orthoclase or microcline, and is formed by the recrystallization of the original potassic feldspar contained in the rocks. In the granitic ores of the Stratton's Independence, Portland, Ajax, and Elkton mines, this secondary feldspar is the principal gangue mineral.

Oxidized ores, while still worked in many properties, are of relatively less importance than when Penrose described the district. They contain the characteristic dull gold, often in pseudomorphous skeletons, resulting from the oxidation of the tellurides, associated with tellurite (tellurium dioxide), emmonsite or durdenite (both hydrated ferric tellurites), and probably other oxidized compounds of tellurium and iron. These minerals occur in association with kaolin, alunite, and ferruginous clays. The deep workings of the present day show that kaolin is always connected with oxidation, and is not a product of the original mineralization of the district, as was supposed by Penrose.

The Cripple Creek ores, as a rule, contain very little silver, the average proportion being about 1 ounce of silver to 10 ounces of gold. In the Portland and Stratton's Independence mines the proportion is very much less, the silver from the Portland in 1901 amounting to only 2.4 ounces for each 100 ounces of gold. In the Blue Bird, Doctor-Jack Pot, Conundrum, Pointer, and other mines containing notable amounts of tetrahedrite or galena, the proportion of silver rises considerably above the average.

The average value of the Cripple Creek ores lies probably between $30 and $40 per ton. In some of the larger mines the average value sinks to about $25 per ton. From a lower economic limit of about $12 per ton the values of individual shipments swing through a wide range up to ores carrying $3,000 or $4,000, or even $8,000, per ton. Occasionally smaller amounts—one or two tons—have yielded as much as $50,000 per ton.

STRUCTURAL CHARACTER OF DEPOSITS.

With few exceptions the ore bodies, of whatever shape, are causally connected with fissures, and most of them constitute fissure veins of various types. The fissure system of the district appears to radiate from a point near the northern limit of the volcanic area. In the eastern part the prevailing directions are northwest or north-northwest, gradually changing to a northerly strike in the southern portion and to predominant north-northeast or northeast courses in the western side of the district.

Individual veins are rarely over half a mile in length, but linked-vein systems often extend for a mile in the same direction. The dip is generally very steep. The movement along these fissure planes appears in all cases to have been very slight. The fissures charged with ore are sometimes simple veins with one fracture plane; much more commonly, however, they are composite veins or lodes which consist of several closely spaced and frequently linked fissures, all more or less ore bearing. A better expression for this structural type as it appears in Cripple Creek is the term "sheeted zone."

TYPES OF DEPOSITS.

The most important types of auriferous ore bodies occurring in the district are:

1. Tabular in form and strictly following simple fissures or sheeted zones. A subtype comprises lodes in which the sheeted zone follows "basalt" or phonolite dikes.

2. Irregular bodies adjacent to fissures and formed by replacement and recrystallization of the country rock—usually granite.

These types are not always sharply distinct, but may be connected by deposits of intermediate character.

All the ore bodies, of whatever type, exhibit certain common features which serve to distinguish the deposits of Cripple Creek from those of most other mining districts. In the first place, the actual openings in the rocks available for the deposition of ore are, as a rule, remarkably narrow. In the second place, the amount of material carried in the mineralizing solutions and deposited as gangue and ore minerals was comparatively small. In consequence of these two conditions, the district contains no such massive veins, solidly filled with quartz or other vein minerals, as are characteristic of the San Juan region in Colorado or the Mother Lode region in California. Even the small fissures of the Cripple Creek district are rarely completely filled, but exhibit a characteristic open or vuggy structure. Where the fractures are of unusual width, or where the rocks are extensively shattered, as in the Midget and Moose mines, the small volume of available vein matter is particularly noticeable. The walls of such fractures and the fragments of the shattered rock are usually merely coated with a thin deposit of quartz, fluorite, and other minerals. As the rich tellurides were usually among the minerals last to form, and are particularly abundant on the walls of the vugs, it is probable that had quartz, fluorite, or other gangue minerals been more abundantly deposited the ores would have been of much lower grade.

Sheeted veins.—The mineralized sheeted zones constitute the most characteristic deposits of the district and occur in practically all the rocks, although particularly common in breccia. They consist of a varying number of narrow, approximately parallel fissures, together composing a sheeted zone that may range from a fraction of a foot to 50 or 60 feet in width. Such uncommonly wide zones of fissuring, however, can usually be resolved into two or more sheeted zones lying so close together that the whole constitutes for practical purposes a single ore body, as in the Captain vein system of the Portland mine. Usually the sheeted zones are from 2 to 10 feet in width. In other cases the fissures may be very numerous, the rock for a foot or more in width being divided into thin parallel slabs, while on each side of this medial portion the fissures become farther and farther apart as the lode grades into the normal country rock of the vicinity. In still other cases there may be two main fissures, 3 or 4 feet apart, accompanied by more or less irregular fracturing of the intervening and adjacent rock. As a rule the fissures are mere cracks, showing no brecciation, slickensiding, or other evidence of tangential movement of the walls. Usually the tellurides are exclusively confined to the narrow fissures and cracks, and do not, in this type of deposit, in any sense constitute a replacement of the country rock. The rocks in the vicinity of the fissures are partly replaced by dolomite, pyrite, and a little fluorite; the telluride ores, however, do not share this propensity, but coat the open fissures, associated with a little quartz and fluorite. Replacement by tellurides does occur in two other types of deposits, to be described later; but as regards the simple veins and sheeted zones, it will be necessary to modify the results of Penrose by restricting the metasomatic rôle of the tellurides. In the oxidized parts of the veins, such as were almost exclusively available for observation when Penrose visited the district, these relations can seldom be clearly ascertained, and would easily lead one to overemphasize replacement as a feature of vein formation. The fissures are not, in general, planes of faulting. Appreciable movement has undoubtedly occurred in some instances, but the displacement probably rarely exceeded 1 or 2 feet.

Although found most abundantly in the breccia or trachytic phonolite, sheeted zones and single fissures are often well developed in the granite, as in the El Paso, C. K. & N., and Gold Coin mines. While in some of these lodes the ore minerals are as plainly confined to the fissures as in the breccia, in other cases the ore to some extent permeates the granite alongside the fissure, this constituting a deposit intermediate in nature between types 1 and 2. They also frequently follow phonolite dikes, the general tendency of these dikes to develop a platy parting parallel to their walls being particularly favorable to the production of a well-defined sheeted zone when the direction of fissuring happens to coincide with that of the dike.

The metasomatic alteration accompanying these sheeted zones is surprisingly slight, and consists of a partial replacement of the breccia, phonolite, trachytic phonolite, or "basalt" by dolomite and pyrite accompanied by a small amount of sericite and a little secondary potash feldspar. But even in the most altered rock the newly formed minerals rarely form more than a small percentage of the rock mass. The alteration in granite exhibits a somewhat different phase, described in a subsequent paragraph.

Not all the sheeted zones carry ore, nor is the ore of a productive sheeted zone necessarily coextensive with the fissuring. The ore occurs in pay shoots up to 2,000 feet in length and 1,000 feet in depth, but usually very much smaller than is indicated by these limits. The boundary between the ore and the barren portions of the lode can be determined, as a rule, only by assays. No single factor that can account for the localization of the ore in these pay shoots has been discovered. In some mines the pay shoots occur where the lode is intersected by cross fissures; in other mines no such relation exists. In some mines ore occurs where the fissures pass through phonolite dikes; in other mines the lode, elsewhere productive, becomes barren when it enters phonolite; while in still others the presence of the phonolite has had no apparent influence upon ore deposition. It thus appears that the occurrence of two or more favorable factors is necessary to determine the position of a pay shoot in a lode. The discovery of these factors is one of the unsolved problems connected with the Cripple Creek district.

Replacement deposits in granite.—The replacement deposits in granite all occur in close proximity to the contact with the breccia, and are well developed in the Elkton (Thompson), Ajax, Independence, and Portland mines. Although these bodies of ore are related to fissures and occur particularly where several fissures intersect, or where they meet a dike, the ore is not confined to the actual fractures. The rock in the vicinity of these fissures is often extensively altered. The change from altered to unaltered rock, while never perfectly sharp, is often fairly abrupt and may take place within a distance of a few feet. The most obvious characteristic of the metamorphosed rock is a porous texture and a change of the reddish color of the normal granite to grayish or greenish tints. Closer examination shows that, while the porphyritic aggregates of pink microcline, so prominent in the Pikes

Peak type of granite, may remain unaltered, the rest of the rock, consisting originally of microcline, oligoclase, quartz, and biotite, may be completely recrystallized as a porous, vuggy aggregate of secondary orthoclase (valencianite), quartz, fluorite, pyrite, calaverite or sylvanite, and, in exceptional cases, sphalerite and galena. The ore minerals are partly inclosed in the other secondary minerals, but occur most abundantly with little projecting crystals of fluorite, quartz, and valencianite on the walls of the irregular pores so characteristic of the altered rock. The biotite of the original granite yields most readily to alteration, and, in rock otherwise almost entirely unaltered, may be changed to an aggregate of fluorite, quartz, and ore minerals. Some of the ore of the Ajax mine exhibits well this initial stage of alteration. With further alteration the original quartz and oligoclase of the granite are attacked. The quartz, originally in large homogeneous and irregular grains, recrystallizes as aggregates. Secondary orthoclase or valencianite forms often in clear, sharply idiomorphic crystals, which either project into open cavities or form aggregates with the secondary quartz. In many cases, however, the secondary valencianite results from the recrystallization of the older microcline practically in place. The two generations are sometimes distinguishable by the greater clearness and more or less idiomorphic form of the younger mineral and the absence of the characteristic microcline twinning. But it is often impossible to determine the line between feldspar which, from its association with quartz and fluorite, is clearly secondary and the original microcline of the granite. Occasionally a little calcite may be detected in the altered granite, but this is rare. The original apatite and zircon of the granite are not, so far as observed, affected by the alteration described.

While the replacement deposits in granite are important because of their size and the readiness with which the ore may be mined free from waste, the ore itself is usually of lower grade than that formed in the fissures of the sheeted zones.

Mineralized "basalt" dikes.—The ore bodies formed by the mineralization of basic dikes are in some ways closely related to the sheeted zones already described. Like the phonolite dikes, the "basalt" exhibits a pronounced tendency to split into thin sheets parallel with the dike walls. Normally, the minute fissures so formed are filled with veinlets of calcite and contain no ore. When, however, a zone of fissuring coincides with the dike the latter may be traversed by veinlets of quartz and fluorite carrying sylvanite or calaverite, while the body of the dike may be impregnated with pyrite. Such ore differs from that of the usual sheeted zones in breccia or phonolite in that the tellurides are not so clearly confined to the actual fissures, but appear to some extent to permeate the rock with the pyrite. The richest portion of the ore, however, undoubtedly occurs in the small veinlets in the dike, and usually near one or both walls where the fissuring is best developed. The occurrence of rich ore bodies in basic dikes in the Portland (Anna Lee), Moose, Elkton, Conundrum, Pinto, and other mines has tended to exaggerate the importance of these dikes in general, and some have even supposed a genetic relation to exist between them and the mineralization of the district.[a] Such an hypothesis, however, loses sight of the vast amount of profitless work that has been expended in the district in driving on the usually unproductive basaltic dikes, and the very small proportion of the known ore bodies that can be shown to have any connection whatever with these intrusions. That the basaltic dikes are not always readily mineralized even when accompanied by fissuring is shown in the interesting case of the Strong mine, where the ore occurs as mineralized granite on each side of the dike, while the latter is barren.

DEPTH OF OXIDIZED ZONE.

At a few points, as in the Abe Lincoln and El Paso mines, tellurides are found almost at the surface. It is much more common, however, to find an upper zone, from 200 to 400 feet deep, in which free gold prevails and which gradually changes to the zone of pure telluride ores. As may be expected from the varying surface form and conditions of drainage, there is great range in the depth attained by oxidation. Partial oxidation extends in many mines to a depth of over 1,000 feet, especially along the often more or less open fissures. In the Wild Horse mine the zone of complete oxidation reaches a depth of 1,100 feet and then suddenly ends. In the Isabella mine partial oxidation attained at least 1,200 feet, and the same applies to the Gold Coin mine in Victor, although telluride ores prevail at that depth as well as in many levels above. The question is chiefly one of depth of ground water and of facilities for circulation of oxygen. Further data bearing upon this problem may be found on page 31.

RELATIONS OF ORE BODIES TO DEPTH.

It is well known that the payable ores in auriferous lodes are rarely equally distributed in the lode, but form tabular bodies of more or less regular outline. The projections of these ore bodies on the plane of the lode often appear as elongated areas with greater vertical than horizontal extent. The ore bodies or shoots of Cripple Creek show great similarity to those of other gold-bearing veins; their limit in depth is usually as well defined as their extent in a horizontal direction.

Of sixty pay shoots of Cripple Creek mines plotted together for

---

[a] Stevens, E. A., Basaltic zones as guides to ore deposits in the Cripple Creek district: Trans. Am. Inst. Min. Eng., vol. 33, 1903, p. 686.

purposes of comparison, over thirty extended from the surface to a depth of less than 500 feet. The maximum individual production of these is less than $1,000,000. Near six of these ore bodies further exploration developed new shoots below the old ones, but usually of smaller extent. In practically all thirty cases the development work had been carried down a few hundred feet below the last ore of the surface shoot. The form of these smaller shoots is often equidimensional; in a few cases the horizontal extent is greater than the vertical, or the shoot is wholly irregular; in many cases the shoot pitches steeply northward on the plane of the vein and the ratio of vertical to horizontal extent is 2 : 1 or 3 : 1.

In eight of the sixty cases the shoot extended from the surface to a depth of 1,000 feet, or a little more, and ended. Further development to about 1,500 feet failed to find new shoots of any importance, though small pockets were often discovered. In six of these eight cases the ratio of vertical to horizontal extent varies from 3 : 1 to 5 : 1, and the shoots usually pitch northward at angles of 60° to nearly 90° from the horizontal. In the remaining two cases the shoots have about the same horizontal as vertical extent. The maximum horizontal length is 1,300 feet, while 400 is much more common. · In two of the sixty cases the pay shoot is 1,500 to 2,000 feet long, maximum depths of 600 and 1,000 feet having been attained and the bottom level being still in ore. In thirteen of the sixty cases the shoot began over 200 feet below the surface; in eight of these the bottom of the shoot has been reached, while in five the lowest level is still in ore. Steeply dipping, irregular elongated forms prevail. Many of this group of thirteen represent veins parallel and close to those on which pay shoots outcropping at the surface were found.

These statements will give an idea of the form of the shoots. Of course, in the case of shoots reaching the surface, a certain part has probably been removed by erosion. Judging from the shoots which distinctly began below the surface, the normal form of the ore bodies is elongated, vertical, or pitching sharply northward, the ratio of vertical to horizontal extension varying from $1\frac{1}{2}$: 1 to 5 : 1. Some of these shoots are, however, of about equal dimensions, vertically and horizontally, while in a few the horizontal dimension is the greater.

Of the known ore bodies, as few exceed 1,000 feet in length, so very few exceed 1,000 feet in depth or extend more than 1,000 feet from the surface. Speaking broadly, explorations below that limit have not proved very satisfactory. Drawing the lines a little closer, it may be said that in proportion to the amount of exploration the upper 700 or 800 feet have yielded more than the interval from that limit to the lowest levels reached—about 1,500 feet. It must not be overlooked, however, that four or five mines still have good ore bodies at a depth of 1,200 to 1,400 feet from the surface. The developments of the next year or two will probably give a safer basis for generalization.

Roughly speaking, the above-mentioned distribution holds good for any elevation within the district. In other words, the principal productive zone everywhere occupies the space from the surface down to about 1,000 feet below it, and its lower limit thus forms a curved surface approximately parallel to the surface of the ground.

It is probable that the minimum depth of rock removed from the district by erosion amounts to 1,000 feet in the central part and to 400 or 500 feet about the periphery. The shape and number of the ore bodies formerly existing in this eroded zone can be only conjectured. It is probable that the veins were formed shortly after the close of igneous activity, while the volcano yet possessed a much greater height than at present. The absence of hot waters and the depth of oxidation attained indicate that vein formation at Cripple Creek is by no means a recent phenomenon.

The general features of the vertical distribution of the known ore bodies recorded above have of late years received more or less recogultion, and there has been a decided tendency to attribute them to a process of secondary enrichment effected by waters moving generally downward from the surface. It has been supposed [a] that such waters have carried down a part of the auriferous contents of those portions of the lodes now removed by erosion and have enriched originally lean pyritic ores by the secondary deposition of gold and silver tellurides and argentiferous tetrahedrite, with associated gangue minerals.

It is clear that the hypothesis in question is suggested by the distribution of known pay shoots. The question arises, How far does the distribution of known pay shoots represent the distribution of all the pay shoots in the district? In other words, How far has exploration been impartial in revealing ore bodies near the surface and at depths greater than 1,000 feet?

It requires but little examination to make clear the fact that ore bodies within 1,000 feet of the surface are far more likely to be discovered than those at greater depth. While shafts have been sunk for a few hundred feet without any indication of ore and have ultimately been developed into productive mines, such a procedure is considered bold prospecting, and few well-informed mining men would seriously contemplate sinking a shaft over 1,000 feet in depth solely on the expectation of finding possible ore bodies below that depth. Most of the large mines in the district have started upon some indication of ore near the surface and have grown by the subsequent discovery of other lodes and ore bodies in the course of their underground development. As few individual ore bodies persist for more than 1,000 feet in depth, by

---

[a] Bancroft, Geo. J., Eng. and Min. Jour., vol. 74, 1902, pp. 752–753, and vol. 75, 1903, pp. 111–112.
Finch, J. W., Proc. Colorado Sci. Soc., vol. 7, 1904, pp. 193–252.

far the greater part of the underground prospecting is at less depths, there being usually little inducement to go deeper, unless, as in the case of the Gold Coin and Portland mines, lodes are discovered in which the ore, beginning several hundred feet below the surface, extends deeper than the pay shoot upon which the mine was originally opened. · Thus deep prospecting is usually confined to the vicinity of the larger and more persistent pay shoots which have been followed down from near the surface. Underground water has also proved a most serious obstacle to deep prospecting, few properties being able to develop. below the 1,000-foot zone unless there is abundant and high-grade ore in sight.

It may thus be concluded, without necessarily advocating promiscuous exploration below the 1,000-foot zone, that any ore bodies existing below that depth are far less likely to be discovered than those above, where from the surface to depths of several hundred feet the rocks of the district are riddled with shafts, drifts, crosscuts, and adits. It is exceedingly difficult, however, to determine, even approximately, the relative importance of this factor in the problem. It is probably safe to assume that the chances of discovering a given ore body within the 1,000-foot zone are at least ten times those of discovering an ore body below that zone, and the ratio may be very much greater. It is probably true that there was originally more ore within the 1,000-foot zone than there is in a corresponding zone below, but this disparity is not necessarily anything like so great as is indicated by the vertical distribution of known pay shoots.

Another important line of inquiry bearing upon the relations of the ore bodies to depth is concerned with the question of the relative size and abundance of the fissures near the surface and at greater depth. It has been shown that all the ore bodies are intimately connected with fissures. If such fissures are generally smaller and less abundant below the 1,000-foot zone than they are within it, obviously there is introduced a factor which diminishes the supposed importance of secondary enrichment by affording an anterior and physical explanation for the decrease of ore with increase of depth.

Detailed examination of practically all the accessible mines in the Cripple Creek district has led to the conclusion that the fissures, which ordinarily are narrow and often appear as mere cracks, do become less abundant and less conspicuous as greater depth is attained. No mine exhibits this feature better than the Stratton's Independence, in which the very complex systems of productive fissures on the fifth and higher levels contrast most strikingly with the few, insignificant, and nnproductive fractures visible on the fourteenth level. In less degree the same feature is shown in many others of the deep mines, but the rule is not without some very marked exceptions.

The dependence of the ore zone on the surface would then merely express the depth to which fissuring extended in a conical volcanic mountain.

We have thus two factors of importance to account for the scarcity of ore shoots below the 1,000-foot level—first, difficulties of development and exploration, and second, the disappearance of fissures in depth. They do not seem to be sufficient, however, and it is believed that a third factor, as yet undiscovered, exists, and that it is related to the chemistry of the actual ore deposition.

In those districts where so-called secondary sulphide enrichment is known to have taken place the ore minerals exhibit in general an orderly sequence, both in relative abundance and in kind, from those characteristic of the most highly enriched ore near the zone of oxidation to those constituting the original, lean, and unaltered ore. The secondary minerals produced are such as can result from rearrangement and concentration of elements present in different combinations in the primary ores. At certain points within this range of alteration it is possible to detect direct mineralogical evidence of the change of one mineral to another, effected by solutions moving downward from the zone of oxidation. In most cases the secondarily enriched ores bear a recognizable relation to the lower limit of oxidation.

Careful study of the Cripple Creek ore deposits has failed to discover that the hypothesis of secondary enrichment is supported by crucial evidence of the kind just indicated. The minerals are not arranged in any discoverable definite sequence, nor does the present investigation find much to support the view that the rich telluride ores, as a rule, pass with increasing depth into low-grade pyritic ores. Frequently such ore as occurs below a depth of 1,000 feet is precisely the same in character as ore found within 100 feet of the surface. Tetrahedrite, which has been regarded by some, without definite proof, as a secondary mineral, occurs sporadically throughout the district and at all depths reached by present workings. The richest ore does not uniformly occur immediately below the oxidized ore. There is, in fact, little indication of enrichment in the oxidized zone such as is so often found in gold-quartz veins of the normal type. Frequently the fresh telluride ore is extremely rich, and high-grade pockets occur impartially in oxidized and fresh portions of the veins. Neither would it be correct to say that there is a gradual decrease in the value of ore in depth. It is quantity, not value, which decreases.

While it is certain that pyrite, and possibly other minerals, has formed at more than one period during the mineralization of the district, and while it is equally clear that in general the rich tellurides were the last of the ore minerals to be deposited, there is apparently no evidence that any one of these minerals has been formed by enriching solutions at the expense of primary minerals. So far as definite conclusion is warranted in an investigation as yet incomplete, it appears that the unoxidized ore deposits of the Cripple Creek district represent the product of one general period of mineralization and that they have not been appreciably modified by secondary enrichment during the subsequent erosion of the region.

UNDERGROUND WATER.

The conditions of underground waters are interesting and somewhat unusual. A dry climate and a heavy percentage of run-off minimize the annual additions to the underground supply. Nevertheless, the ground-water level is not unusually deep, and large quantities of water are encountered in all the mines below that level. The original water surface of the district in the volcanic rocks stood at elevations of 9,400 to 9,700 feet, or 100 to 600 feet below the surface of the ground. At first pumping was commenced by individual mines, but it was soon found that the radius of drainage had unusual length—that is, that one mine would drain others situated at a distance. Drainage tunnels were then undertaken, and the Chicago and Cripple Creek, the Ophelia, the Standard, and lately the El Paso tunnels were driven, each of which practically accomplished the drainage of a large part of the district almost down to its own level, thus showing that the ground water is limited in quantity and is more of the nature of a local reservoir than a "subterranean sea."

The plug of volcanic rocks which fills the throat of the old volcano is rudely circular, with a diameter of 3 miles. This mass is extremely porous, and is, moreover, cut in many directions by partly filled fissures and sheeted zones, so that water can circulate within it with comparative freedom in several directions. It retains this character down to the greatest depth yet reached. On the other hand, the surrounding granite is relatively impermeable and is less traversed by open fissures. No doubt it contains ground water down to a depth of 2,000 feet or more, but in very much smaller quantity, and the circulation of this water must be extremely slow. This is clearly shown by the fact that the water in the breccia is not drained by Cripple Creek and Arequa Gulch below the level of the points where they leave the volcanic area. Thus the volcanic plug resembles a water-soaked sponge inserted in a hole cut in an impermeable substance. The drainage of the mines is thereby greatly facilitated, as it is not necessary to extend the tapping tunnels to each mine.

The El Paso tunnel, completed in the winter of 1903–4, has an elevation of 8,790 feet at the portal. Within a short time it effectually drained not only the Beacon Hill mines but also the Gold Hill mines, and its influence extended even to the Last Dollar and the Elkton mines. But the foregoing statement in relation to draining the distriet must be so modified as to exclude a certain part on the eastern side, comprising the mines about Independence on the east side of Bull Hill and those on Battle Mountain and in the town of Victor, in which the effect of the El Paso tunnel is slight. The Findley, Hull City, Vindicator, and Golden Cycle mines about the town of Independence seem to occupy a separate drainage basin, probably divided from the main area by masses of relatively impermeable rock.

The Portland, Stratton's Independence, and the other mines near Victor occupy another drainage basin. Of these the Gold Coin and the Stratton's Independence have shafts below the level of the El Paso tunnel, and their pumps have probably drained the surrounding territory to a considerable extent. The influence of the drainage tunnel on the Portland mine is a question upon which opinions differ.

The subject of the drainage of the Cripple Creek mines has been actively studied in late years, and valuable contributions to the subject have been made by Mr. Victor G. Hills and others, who have shown that the water can be tapped, without prohibitive expense, down to an elevation of 7,500 feet, or 1,300 feet below the El Paso tunnel; for this depth a tunnel $3\frac{1}{2}$ miles long would be needed. On account of the great porosity of the rocks it is not probable that the next thousand feet below the El Paso tunnel level will show any great diminution in the amount of water stored in the rocks.

SUBTERRANEAN GASES.

During the earlier years of Cripple Creek no unusual amount of mine gases was observed, but, as the shafts and workings deepened, several properties began to experience much annoyance and even serious interference with work, often in spite of vigorous measures for ventilation. These gases appear to issue chiefly from the breccia, especially where it is of porous and loose texture, but they sometimes flow from partly open vein fissures in such quantity that a light held up to the fissures is immediately extinguished. Thus far the mines on Battle Mountain and those of the Golden Cycle, Vindicator, and Isabella groups have suffered no inconvenience, but most of the mines west of this line have had more or less trouble with this subtle and insidious enemy. At least one mine has been forced to close down entirely, and several others are often obliged to abandon work for days. Sometimes the amount of gas issuing is small, and ordinary ventilation will carry it away; again it may issue in large volumes and practically fill the mine for some time. In other mines the gas persistently hangs at certain places, forming barriers which can be passed only with difficulty. The outflow of gas is unquestionably related to barometric fluctuations, though it is usually locally reported to be influenced by the direction of the wind. The investigations concerning this subject are not yet concluded. It may be said, however, that a sudden lowering of the barometer seems to be the most important factor. Upon such a fall the gas issues in great volumes, but decreases when the barometer remains steadily low for several days. The gas is often very heavy, filling lower parts of drifts and winzes like water, and cases are reported in which it has actually been bailed from a shaft. Its temperature is somewhat higher than that prevailing in the mine under normal conditions.

It has practically no smell or taste, but small quantities of it easily produce effects of suffocation. Miners working in places where this gas is mixed with the air soon experience various forms of physical distress, and several fatal accidents have been caused by men entering drifts and winzes filled with it.

The characteristics of the gas seemed to point to carbon dioxide, and it is generally so termed. Preliminary determinations of carbon dioxide by a portable apparatus yielded percentages which seemed far too small in comparison with the effects of the gas examined, and led to the belief that some other substance was present. Samples were then collected and analyzed. The analyses showed the gas to be a mixture of nitrogen with about 20 per cent carbon dioxide and a small amount of oxygen.

The occurrence of these exhalations over a large part of the ore-bearing area is of much interest. They certainly increase in quantity with depth, and it is to be feared that in some cases they may seriously affect mining operations. The evil has proved very difficult to cope with. Ventilation alone has rarely proved efficient, and the only practicable remedial measures appear to be cementation of drifts at particularly bad places and working the mine under air lock at a pressure slightly exceeding the normal.

The origin of these gases can not reasonably be sought in any such explanation as the oxidizing of sulphides and accompanying absorption of oxygen. We believe that they represent the last exhalations from the throat of the extinct Cripple Creek volcano.

FUTURE OF THE DISTRICT.

To predict the future yield of any mining district is no easy task; the conditions under which most ores are deposited are as yet too imperfectly understood, and the deposits themselves are usually too erratic in form and distribution, to give certitude to such predictions, even when these are based upon a careful study of the history and present condition of a district. Nevertheless, it is part of the duty of the geologists who have officially investigated the Cripple Creek district to interpret to the best of their ability the bearing of ascertained facts upon future mining development. For such a forecast of the future moderate probability is all that can be claimed.

As has been pointed out in the preceding pages, the largest known ore bodies of the district are apparantly confined within a zone which extends from the surface to a depth of 1,000 feet. In general, explorations below that depth have been much less satisfactory, as regards quantity of ore, than explorations above. It is certainly true that some large ore bodies as yet show no sign of depletion in depth, and that some good pay shoots have been found at a depth of 1,400 to 1,500 feet. On the other hand, the number of ore shoots that have been exhausted with increase in depth is considerable.

It is probable that the ore bodies, known or unknown, occurring below the 1,000-foot zone are neither so large nor so abundant as those nearer the surface. The discovery and exploitation of these deeper ore bodies is, moreover, beset with increasing difficulties, chief among which is the problem of dealing with the underground water. For these reasons it is unlikely that the zone between the 1,000-foot and 2,000-foot levels will yield as much as the zone between the surface and the 1,000-foot level, but the possibility is not denied that some strong fissures may carry payable ore to far greater depths than those yet attained.

As regards the zone above the 1,000-foot or 1,500-foot level, it is well to bear in mind that it still contains much ore, both as parts of known ore shoots and as yet undiscovered ore bodies. It is certain that many of these undeveloped ore bodies will be mined in the near future and that this zone will contribute the most important part of the production.

It is probable that the production of the district, while exhibiting fluctuations, will on the whole slowly decline. New ore bodies will undoubtedly be discovered from time to time, and individual mines may be as profitable in the future as they have been in the past, or even more profitable. An increased output may be expected to follow each successful step in deep drainage. But existing conditions indicate that if the maximum production of $18,000,000, in 1900, is to be surpassed the increase will be due to the ore bodies encountered in the upper zone.

INDEX.

35

O

PUBLICATIONS OF UNITED STATES GEOLOGICAL SURVEY.

[Bulletin No. 254.]

The serial publications of the United States Geological Survey consist of (1) Annual Reports, (2) Monographs, (3) Professional Papers, (4) *Bulletins, (5) Mineral Resources, (6) Water-Supply and Irrigation Papers, (7) Topographic Atlas of United States—folios and separate sheets thereof, (8) Geologic Atlas of the United States—folios thereof. The classes numbered 2, 7, and 8 are sold at cost of publication; the others are distributed free. A circular giving complete lists may be had on application.

The Professional Papers, Bulletins, and Water-Supply Papers treat of a variety of subjects, and the total number issued is large. They have therefore been classified into the following series: A, Economic geology; B, Descriptive geology; C, Systematic geology and paleontology; D, Petrography and mineralogy; E, Chemistry and physics; F, Geography; G, Miscellaneous; H, Forestry; I, Irrigation; J, Water storage; K, Pumping water; L, Quality of water; M, General hydrographic investigations; N, Water power; O, Underground waters; P, Hydrographic progress reports. This bulletin is the forty-ninth in Series A and the sixty-first in Series B, the complete lists of which follow. (PP=Professional Paper; B=Bulletin; WS=Water-Supply Paper.)

SERIES A, ECONOMIC GEOLOGY.

B 21. Lignites of Great Sioux Reservation: Report on region between Grand and Moreau rivers, Dakota, by Bailey Willis. 1885. 16 pp., 5 pls. (Out of stock.)

B 46. Nature and origin of deposits of phosphate of lime, by R. A. F. Penrose, jr., with introduction by N. S. Shaler. 1888. 143 pp. (Out of stock.)

B 65. Stratigraphy of the bituminous coal field of Pennsylvania, Ohio, and West Virginia, by I. C. White. 1891. 212 pp., 11 pls. (Out of stock.)

B 111. Geology of Big Stone Gap coal field of Virginia and Kentucky, by M. R. Campbell. 1893. 106 pp., 6 pls.

B 132. The disseminated lead ores of southeastern Missouri, by Arthur Winslow. 1896. 31 pp.

B 138. Artesian-well prospects in Atlantic Coastal Plain region, by N. H. Darton. 1896. 228 pp., 19 pls. (Out of stock.)

B 139. Geology of Castle Mountain mining district, Montana, by W. H. Weed and L. V. Pirsson. 1896. 164 pp., 17 pls.

B 143. Bibliography of clays and the ceramic arts, by I. C. Branner. 1896. 114 pp.

B 164. Reconnaissance on the Rio Grande coal fields of Texas, by T. W. Vaughan, including a report on igneous rocks from the San Carlos coal field, by E. C. E. Lord. 1900. 100 pp., 11 pls.

B 178. El Paso tin deposits, by W. H. Weed. 1901. 15 pp., 1 pl.

B 180. Occurrence and distribution of corundum in United States, by J. H. Pratt. 1901. 98 pp., 14 pls.

B 182. A report on the economic geology of the Silverton quadrangle, Colorado, by F. L. Ransome. 1901. 266 pp., 16 pls.

B 184. Oil and gas fields of the western interior and northern Texas Coal Measures and of the Upper Cretaceous and Tertiary of the western Gulf coast, by G. I. Adams. 1901. 64 pp., 10 pls. (Out of stock.)

B 193. The geological relations and distribution of platinum and associated metals, by J. F. Kemp. 1902. 95 pp., 6 pls. (Out of stock.)

B 198. The Berea grit oil sand in the Cadiz quadrangle, Ohio, by W. T. Griswold. 1902. 43 pp., 1 pl.

PP 1. Preliminary report on the Ketchikan mining district, Alaska, with an introductory sketch of the geology of southeastern Alaska, by Alfred Hulse Brooks. 1902. 120 pp., 2 pls.

B 200. Reconnaissance of the borax deposits of Death Valley and Mohave Desert, by M. R. Campbell. 1902. 23 pp., 1 pl.

B 202. Tests for gold and silver in shales from western Kansas, by Waldemar Lindgren. 1902. 21 pp.

PP 2. Reconnaissance of the northwestern portion of Seward Peninsula, Alaska, by A. J. Collier. 1902. 70 pp., 11 pls.

PP 10. Reconnaissance from Fort Hamlin to Kotzebue Sound, Alaska, by way of Dall, Kanuti, Allen, and Kowak rivers, by W. C. Mendenhall. 1902. 68 pp., 10 pls.

PP 11. Clays of the United States east of the Mississippi River, by Heinrich Ries. 1903. 298 pp., 9 pls.

PP 12. Geology of the Globe copper district, Arizona, by F. L. Ransome. 1903. 168 pp., 27 pls.

B 212. Oil fields of the Texas-Louisiana Gulf Coastal Plain, by C. W. Hayes and William Kennedy. 1903. 174 pp., 11 pls.

B 213. Contributions to economic geology, 1902; S. F. Emmons, C. W. Hayes, geologists in charge. 1903. 449 pp.

PP 15. The mineral resources of the Mount Wrangell district, Alaska, by W. C. Mendenhall and F. C. Schrader. 1903. 71 pp., 10 pls.

B 218. Coal resources of the Yukon, Alaska, by A. J. Collier. 1903. 71 pp., 6 pls.

B 219. The ore deposits of Tonopah, Nevada (preliminary report), by J. E. Spurr. 1903. 31 pp., 1 pl.

PP 20. A reconnaissance in northern Alaska, in 1901, by F. C. Schrader. 1904. 139 pp., 16 pls.

PP 21. Geology and ore deposits of the Bisbee quadrangle, Arizona, by F. L. Ransome. 1904. 168 pp., 29 pls.

B 223. Gypsum deposits of the United States, by G. I. Adams and others. 1904. 129 pp., 21 pls.

PP 24. Zinc and lead deposits of northern Arkansas, by G. I. Adams. 1904. 118 pp., 27 pls.

PP 25. Copper deposits of the Encampment district, Wyoming-Colorado, by A. C. Spencer. 1904. 107 pp., 2 pls.

B 225. Contributions to economic geology, 1903; S. F. Emmons, C. W. Hayes, geologists in charge. 1904. 527 pp., 1 pl.

PP 26. Economic resources of the northern Black Hills, by J. D. Irving, with contributions by S. F. Emmons and T. A. Jaggar, jr. 1904. 222 pp., 20 pls.

PP 27. A geological reconnaissance across the Bitterroot Range and Clearwater Mountains in Montana and Idaho, by Waldemar Lindgren. 1904. 123 pp., 15 pls.

B 229. Tin deposits of the York region, Alaska, by A. J. Collier. 1904. 61 pp., 7 pls.

B 236. The Porcupine placer district, Alaska, by C. W. Wright. 1904. 35 pp., 10 pls.

B 238. Economic geology of the Iola quadrangle, Kansas, by G. I. Adams, Erasmus Haworth, and W. R. Crane. 1904. 83 pp., 11 pls.

B 243. Cement materials and industry of the United States, by E. C. Eckel. 1905. — pp., 15 pls.

B 246. Zinc and lead deposits of northwestern Illinois, by H. Foster Bain. 1904. 56 pp., 5 pls.

B 247. The Fairhaven gold placers, Seward Peninsula, Alaska, by F. H. Moffit. 1905. — pp., 14 pls.

B 249. Limestones of southwestern Pennsylvania, by F. G. Clapp. 1905. — pp., 7 pls.

B 250. The petroleum fields of the Pacific coast of Alaska, with an account of the Bering River coal deposit, by G. C. Martin. 1905. — pp., 7 pls.

B 251. The gold placers of the Fortymile, Birch Creek, and Fairbanks regions, Alaska, by L. M. Prindle. 1905. — pp., 16 pls.

WS 117. The lignite of North Dakota and its relation to irrigation, by F. A. Wilder. 1905. — pp., — pls.

PP 36. The lead, zinc, and fluorspar deposits of western Kentucky, by E. O. Ulrich and W. S. Tangier Smith. 1905. — pp., — pls.

PP 38. Economic geology of the Bingham mining district of Utah, by J. M. Boutwell, with a chapter on areal geology, by Arthur Keith, and an introduction on general geology, by S. F. Emmons. 1905. — pp., — pls.

PP 41. The geology of the central Copper River region, Alaska, by W. C. Mendenhall. 1905. — pp., — pls.

B 254. Report of progress in the geological resurvey of the Cripple Creek district, Colorado, by Waldemar Lindgren and F. L. Ransome. 1904. 36 pp.

SERIES B, DESCRIPTIVE GEOLOGY.

B 23. Observations on the junction between the Eastern sandstone and the Keweenaw series on Keweenaw Point, Lake Superior, by R. D. Irving and T. C. Chamberlin. 1885. 124 pp., 17 pls.

B 33. Notes on geology of northern California, by J. S. Diller. 1886. 23 pp. (Out of stock.)

B 39. The upper beaches and deltas of Glacial Lake Agassiz, by Warren Upham. 1887. 84 pp., 1 pl. (Out of stock.)

B 40. Changes in river courses in Washington Territory due to glaciation, by Bailey Willis. 1887. 10 pp., 4 pls. (Out of stock.)

B 45. The present condition of knowledge of the geology of Texas, by R. T. Hill. 1887. 94 pp. (Out of stock.)

B 53. The geology of Nantucket, by N. S. Shaler. 1889. 55 pp., 10 pls. (Out of stock.)

B 57. A geological reconnaissance in southwestern Kansas, by Robert Hay. 1890. 49 pp., 2 pls.

B 58. The glacial boundary in western Pennsylvania, Ohio, Kentucky, Indiana, and Illinois, by G. F. Wright, with introduction by T. C. Chamberlin. 1890. 112 pp., 8 pls. (Out of stock.)

B 67. The relations of the traps of the Newark system in the New Jersey region, by N. H. Darton. 1890. 82 pp. (Out of stock.)

B 104. Glaciation of the Yellowstone Valley north of the Park, by W. H. Weed. 1893. 41 pp., 4 pls.

B 108. A geological reconnaissance in central Washington, by I. C. Russell. 1893. 108 pp., 12 pls. (Out of stock.)

B 119. A geological reconnaissance in northwest Wyoming, by G. H. Eldridge. 1894. 72 pp., 4 pls.

B 137. The geology of the Fort Riley Military Reservation and vicinity, Kansas, by Robert Hay. 1896. 35 pp., 8 pls.

B 144. The moraines of the Missouri Coteau and their attendant deposits, by J. E. Todd. 1896. 71 pp., 21 pls.

B 158. The moraines of southeastern South Dakota and their attendant deposits, by J. E. Todd. 1899. 171 pp., 27 pls.

B 159. The geology of eastern Berkshire County, Massachusetts, by B. K. Emerson. 1899. 139 pp., 9 pls.

B 165. Contributions to the geology of Maine, by H. S. Williams and H. E. Gregory. 1900 212 pp., 14 pls.

WS 70. Geology and water resources of the Patrick and Goshen Hole quadrangles in eastern Wyoming and western Nebraska, by G. I. Adams. 1902. 50 pp., 11 pls.

B 199. Geology and water resources of the Snake River Plains of Idaho, by I. C. Russell. 1902. 192 pp., 25 pls.

PP 1. Preliminary report on the Ketchikan mining district, Alaska, with an introductory sketch of the geology of southeastern Alaska, by A. H. Brooks. 1902. 120 pp., 2 pls.

PP 2. Reconnaissance of the northwestern portion of Seward Peninsula, Alaska, by A. J. Collier. 1902. 70 pp., 11 pls.

PP 3. Geology and petrography of Crater Lake National Park, by J. S. Diller and H. B. Patton. 1902. 167 pp., 19 pls.

PP 10. Reconnaissance from Fort Hamlin to Kotzebue Sound, Alaska, by way of Dall, Kanuti, Allen, and Kowak rivers, by W. C. Mendenhall. 1902. 68 pp., 10 pls.

PP 11. Clays of the United States east of the Mississippi River, by Heinrich Ries. 1903. 298 pp., 9 pls.

PP 12. Geology of the Globe copper district, Arizona, by F. L. Ransome. 1903. 168 pp., 27 pls.

PP 13. Drainage modifications in southeastern Ohio and adjacent parts of West Virginia and Kentucky, by W. G. Tight. 1903. 111 pp., 17 pls.

B 208. Descriptive geology of Nevada south of the fortieth parallel and adjacent portions of California, by J. E. Spurr. 1903. 229 pp., 8 pls.

B 209. Geology of Ascutney Mountain, Vermont, by R. A. Daly. 1903. 122 pp., 7 pls.

WS 78. Preliminary report on artesian basins in southwestern Idaho and southeastern Oregon, by I. C. Russell. 1903. 51 pp., 2 pls.

PP 15. Mineral resources of the Mount Wrangell district, Alaska, by W. C. Mendenhall and F. C. Schrader. 1903. 71 pp., 10 pls.

PP 17. Preliminary report on the geology and water resources of Nebraska west of the one hundred and third meridian, by N. H. Darton. 1903. 69 pp., 43 pls.

B 217. Notes on the geology of southwestern Idaho and southeastern Oregon, by I. C. Russell. 1903. 83 pp., 18 pls.

B 219. The ore deposits of Tonopah, Nevada (preliminary report), by J. E. Spurr. 1903. 31 pp., 1 pl.

PP 20. A reconnaissance in northern Alaska in 1901, by F. C. Schrader. 1904. 139 pp., 16 pls.

PP 21. The geology and ore deposits of the Bisbee quadrangle, Arizona, by F. L. Ransome. 1904. 168 pp., 29 pls.

WS 90. Geology and water resources of part of the lower James River Valley, South Dakota, by J. E. Todd and C. M. Hall. 1904. 47 pp., 23 pls.

PP 25. The copper deposits of the Encampment district, Wyoming, by A. C. Spencer. 1904. 107 pp., 2 pls.

PP 26. Economic resources of northern Black Hills, by J. D. Irving, with chapters by S. P. Emmons and T. A. Jaggar, jr. 1904. 222 pp., 20 pls.

PP 27. Geological reconnaissance across the Bitterroot Range and the Clearwater Mountains in Montana and Idaho, by Waldemar Lindgren. 1904. 122 pp., 15 pls.

PP 31. Preliminary report on the geology of the Arbuckle and Wichita mountains in Indian Territory and Oklahoma, by J. A. Taff, with an appendix on reported ore deposits in the Wichita Mountains, by H. F. Bain. 1904. 97 pp., 8 pls.

B 235. A geological reconnaissance across the Cascade Range near the forty-ninth parallel, by G. O. Smith and F. C. Calkins. 1904. 103 pp., 4 pls.

B 236. The Porcupine placer district, Alaska, by C. W. Wright. 1904. 35 pp., 10 pls.

B 237. Igneous rocks of the Highwood Mountains, Montana, by L. V. Pirsson. 1904. 208 pp., 7 pls.

B 238. Economic geology of the Iola quadrangle, Kansas, by G. I. Adams, Erasmus Haworth, and W. R. Crane. 1904. 83 pp., 11 pls.

PP 32. Geology and underground water resources of the central Great Plains, by N. H. Darton. 1905. — pp., 72 pls.

WS 110. Contributions to hydrology of eastern United States, 1904; M. G. Fuller, geologist in charge. 1905. — pp., 5 pls.

B 242. Geology of the Hudson Valley between the Hoosic and the Kinderhook, by T. Nelson Dale. 1904. 63 pp., 3 pls.

PP 34. Delavan lobe of the Lake Michigan glacier of the Wisconsin stage of glaciation and associated phenomena, by W. C. Alden. 1904. 106 pp., 15 pls.

PP 35. Geology of the Perry basin ⋔ southeastern Maine, by G. O. Smith and David White. 1905. — pp., 6 pls.

B 243. Cement materials and industry of the United States, by E. C. Eckel. 1905. — pp., 15 pls.

B 246. Zinc and lead deposits of northeastern Illinois, by H. F. Bain. 1904. 56 pp., 5 pls.

B 247. The Fairhaven gold placers of Seward Peninsula, Alaska, by F. H. Moffit. 1905. — pp., 14 pls.

B 249. Limestones of southwestern Pennsylvania, by F. G. Clapp. 1905. — pp., 7 pls.

B 250. The petroleum fields of the Pacific coast of Alaska, with an account of the Bering River coal deposit, by G. C. Martin. 1905. — pp., 7 pls.

B 251. The gold placers of the Fortymile, Birch Creek, and Fairbanks regions, Alaska, by L. M. Prindle. 1905. — pp., 16 pls.

WS 118. Geology and water resources of a portion of east-central Washington, by F. C. Calkins. 1905. — pp., 4 pls.

B 252. Preliminary report on the geology and water resources of central Oregon, by I. C. Russell. 1905. — pp., 24 pls.

PP 36. The lead, zinc, and fluorspar deposits of western Kentucky, by E. O. Ulrich and W. S. Tangier Smith. 1905. — pp., — pls.

PP 38. Economic geology of the Bingham mining district of Utah, by J. M. Boutwell, with a chapter on areal geology, by Arthur Keith, and an introduction on general geology, by S. F. Emmons. 1905. — pp., — pls.

PP 41. The geology of the central Copper River region, Alaska, by W. C. Mendenhall. 1905. — pp., — pls.

B 254. Report of progress in the geological resurvey of the Cripple Creek district, Colorado, by Waldemar Lindgren and F. L. Ransome. 1904. 36 pp.

Correspondence should be addressed to

THE DIRECTOR,

UNITED STATES GEOLOGICAL SURVEY,

WASHINGTON, D. C.

DECEMBER, 1904.

Author.

Lindgren, Waldemar, 1860-

. . . Report of progress in the geological resurvey of the Cripple Creek district, Colorado; by Waldemar Lindgren and Frederick Leslie Ransome. Washington, Gov't print. off., 1904.

36, iii p. 23½cm. (U. S. Geological survey. Bulletin no. 254.)
Subject series: A, Economic geology, 49; B, Descriptive geology, 61.

1. Geology, Economic—Colorado. I. Ransome, Frederick Leslie, 1868–

Subject.

Lindgren, Waldemar, 1860-

. . . Report of progress in the geological resurvey of the Cripple Creek district, Colorado; by Waldemar Lindgren and Frederick Leslie Ransome. Washington, Gov't print. off., 1904.

36, iii p. 23½cm. (U. S. Geological survey. Bulletin no. 254.)
Subject series: A, Economic geology, 49; B, Descriptive geology, 61.

1. Geology, Economic—Colorado. I. Ransome, Frederick Leslie, 1868–

Series.

U. S. Geological survey.

Bulletins.

no. 254. Lindgren, W. Report of progress in the geological resurvey of the Cripple Creek district, Colo.; by W. Lindgren and F. L. Ransome. 1904.

Reference.

U. S. Dept. of the Interior.

see also

U. S. Geological survey.

Bull. 254—04——4

CPSIA information can be obtained
at www.ICGtesting.com
Printed in the USA
BVHW04*1213180918
527831BV00013B/1018/P